Historical Evidence Concerning Climate Change
Archaeological and Historical Evidence That Man Is Not the Cause

Clayton E. Cramer

Copyright © 2016 Clayton E. Cramer

All rights reserved.

ISBN: 1533073732
ISBN-13: 978-1533073730

Also by Clayton E. Cramer

*My Brother Ron: A Personal and Social History of the
Deinstitutionalization of the Mentally Ill* (2014)

*Armed America: The Remarkable Story of How Guns
Became as American as Apple Pie* (2004)

*Concealed Weapon Laws of the Early Republic: Dueling,
Southern Violence, and Moral Reform* (1999)

Black Demographic Data, 1790–1860: A Sourcebook (1990)

*Firing Back: A Clear, Simple Guide to Defending Your
Constitutional Right to Bear Arms* (1997)

*For the Defense of Themselves and the State:
The Original Intent and Judicial Interpretation of the Right
to Keep and Bear Arms* (1994)

*By The Dim and Flaring Lamps: The Civil War Diary of
Samuel McIlvaine* (1990)

CONTENTS

1 Global Warming? Climate Change? 1
2 Climate Change & Civilization 2
3 Archaeological Evidence 5
4 Not the First Time the Glaciers Are Melting 18
4 Chronicles 29
5 What Does This Mean? 46

ACKNOWLEDGMENTS

I could never have done this without the amazing work of Google digitizing books and journals from places I doubt I could afford to travel.

I have received no funding, assistance, or encouragement from fossil fuel industries to produce this book, but if you are one of those evil institutions, and decide to order 100,000 copies for your stockholders, I would be very grateful!

1 Global Warming? Climate Change?

Over the last twenty years, the theory that mankind's use of fossil fuels is warming up the planet has taken on several names. First it was "global warming." Then, the term "climate change" came into use because the warmists (proponents of the AGW or man-caused climate change theory) claimed that many of the destructive changes would appear in forms that might not at first glance appear as warming, such as increased hurricanes, more severe extremes of temperature and changes in precipitation.

In this book, I am using the term anthropogenic global warming (AGW) for the following reasons: 1) Most Americans (and I suspect many outside the United States) still think in terms of "global warming." 2) I do not dispute that global warming is taking place. My goal is to demonstrate that it is part of an historic sequence predating widespread use of fossil fuels and is therefore not anthropogenic (man-induced). 3) While AGW may well cause significant climate changes above and beyond temperature increases, such changes are outside my field of expertise.

2 Climate Change & Civilization

When I teach Western Civilization at the College of Western Idaho, I always emphasize the role of climate change in the development of human civilization. The end of the Ice Age about 10,000 BC roughly corresponds to the Neolithic Revolution, in which new types of stone tools and the transition from hunter-gatherer societies to widespread agriculture takes place.

Even warmists acknowledge that rising carbon dioxide levels in the atmosphere played a significant part in ending the Ice Age and that the "relatively pleasant global climate of the past 10,000 years is largely thanks to higher levels of atmospheric carbon dioxide."[1]

In historic times, temperature changes have played a significant role in influencing at least European societies. The graph below shows both the change in temperatures and carbon dioxide levels in the atmosphere over the last 2000 years. The years 600-1200 AD are the Medieval Warm Period, a nearly idyllic time when Europe recovered from the political and economic collapse after the fall of Rome. The fall of Rome was inevitable after climate change on the left side of the graph reduced rainfall and started central Asian populations moving in pursuit of more grazing land for their herds. The final step in this series of bumper car collisions westward were the barbarians who pushed over the decaying Roman Empire.[2]

The years from about 1250 or 1300 to 1850 AD are known by historians as the Little Ice Age.[3] The blue line is the center

---

[1] David Biello, "What Thawed the Last Ice Age?", *Scientific American*, April 4, 2012, http://www.scientificamerican.com/article/what-thawed-the-last-ice-age/, last accessed May 3, 2016.

[2] Ellsworth Huntington and Sumner W. Cushing, *Principles of Human Geography*, 2nd ed. (1922), 374.

[3] Brian Fagan, *The Little Ice Age: How Climate Made History, 1300-1850*

point of the temperature reconstruction; the red lines above and below show the 95% confidence interval range.

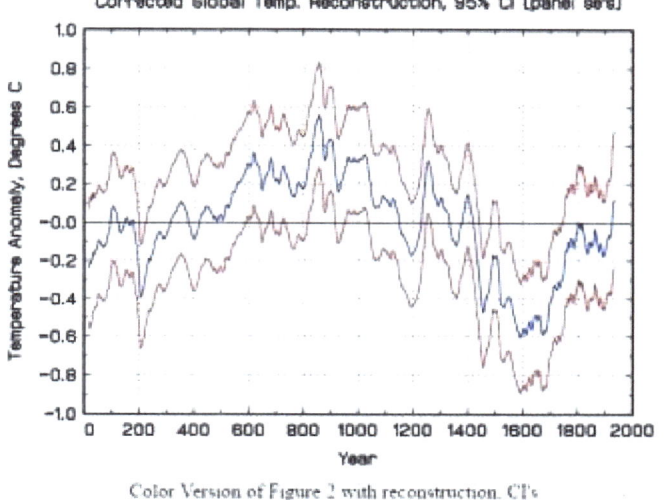

Color Version of Figure 2 with reconstruction CI's   4

Nor is this a modern issue. An ice core at Vostok in Antarctica shows temperature changes over the last 450,000 years. ("Kyr BP" means thousand years before present):

(2001).

[4] The mean relative temperature history of the earth (blue, cool; red, warm) over the past two millennia - adapted from Loehle, C. and McCulloch, J.H. 2008. Correction To: A 2000-Year Global Temperature Reconstruction Based On Non-Tree Ring Proxies Energy & Environment 19:93. Used with permission of C. Loehle.

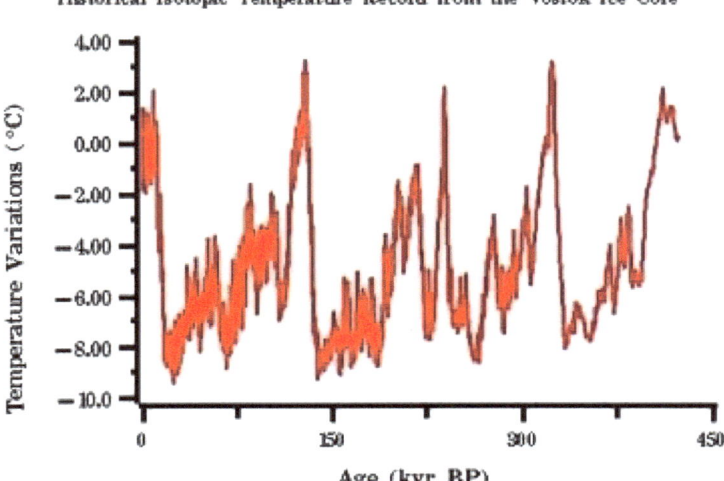

Variation with time of the Vostok isotope temperature record as a difference from the modern surface temperature value of −55.5 °C.

Source: Petit et al.

I think I see a pattern there, every 130,000 years, and I doubt it was caused by mammoths driving SUVs.

This book examine both archaeological and historic evidence concerning climate change. As I explain to my students, historians and archaeologists are both doing the same thing: trying to understand the past. The difference is that historians analyze documents, while archaeologists examine artifacts.

---

[5] J.R. Petit, D. Raynaud, and C. Lorius, "Historical Isotopic Temperature Record from the Vostok Ice Core," http://cdiac.esd.ornl.gov/trends/temp/vostok/jouz_tem.htm, last accessed May 6, 2016.

3 Archaeological Evidence

"The glaciers are melting, the glaciers are melting," is the panicked cry of the warmists. But as the glaciers melt, they reveal evidence of past human habitation at elevations where the glaciers are now, and where they must not have been in the past when humans and their artifacts were deposited. The melting of glaciers, exposing past human habitation has produced a new discipline and journal: *the Journal of Glacial Archaeology*.[6]

One of the recently published works on the subject describes the situation this way:

> As a result of global warming many areas that have been covered with perennial ice are beginning to melt... The melting ice is revealing frozen artifacts and fueling interdisciplinary research. One of the most well-known frozen glacial discoveries was made in 1991 by two hikers in the Alps along the boarder[*sic*] of Italy and Austria... The well-preserved corpse of a 5,000 year old man was found melting out of a glacier at an altitude of 10,500 ft....[7]

Of course if melting ice is exposing remains that must have been at the surface long ago, the ice is hardly perennial, only ancient.

Similarly assuming that the warming is recent:

---

[6] E. James Dixon, M. Callanan, A. Hafner, and P.G. Hare, "The Emergence of Glacial Archaeology," *Journal of Glacial Archaeology* 1:1 (2014).

[7] Jess Anderson-Milhausen, Frozen Organic Artifacts, Museum Practice, and Community Archaeology: An Example from Alaska's Wrangell St. Elias National Park (M.A. thesis, University of Colorado, 2008), 25.

> These environmental changes are resulting in the emergence of artifacts from ancient ice. The discovery of ancient artifacts presents clear and compelling evidence that very old ice is melting for the first time, and climate is changing.[8]

At least this recognized that this ice was ancient not perennial. And again, if you find artifacts in melting ice, it suggests that they were on the surface in the past.

But where? If humans and their artifacts fell into a crevasse or were on the surface of an ancient glacier, and as the glacier moved down slope, they were carried with it. But it seems implausible that our distant ancestors were traipsing around the surface of ancient glaciers.

How common is this discovery of ancient bodies in glaciers? Enough so that there is an entire *book* devoted to it: *Bodies from the Ice*.[9]

Along with glaciers, which move over time, many finds are in ice patches, which do not move and where discoveries are exposed as the ice melts to its old level.

The Yukon is exposing artifacts and remains as the ice patches melt back to their original positions:

> [M]ore than 207 archaeological objects and 1700 faunal remains have been recovered from 43 melting ice patches in the southern Yukon. The artifacts range in age from a 9000-year-old (calendar) dart shaft to a 19th-century musket ball.[10]

---

[8] Linda S. Cordell, Kent Lightfoot, Francis McManamon, George Milner, eds., *Archaeology in America" An Encyclopedia* (2009*)*, 4:302.

[9] James M. Deem, *Bodies from the Ice: Melting Glaciers and the Recovery of the Past* (2008).

[10] P. Gregory Hare, Christian D. Thomas, Timothy N. Topper and Ruth M. Gotthardt, "The Archaeology of Yukon Ice Patches: New Artifacts, Observations, and Insights," 65 Suppl. 1 *Arctic*, 118 (2012).

Melting ice patches in other parts of Canada are also telling a story of old ice melting, revealing signs that they have not always been frozen. A 1991 discovery in the Saint Elias Icefields was "a piece of hide... It had been modified by humans – with slits around the edges and a possible fragment of thong... It appears to have been left by a traveler approximately a thousand years ago." Six years later, a biologist chanced on caribou droppings "a metre and a half deep and the size of a football field, melting from an alpine ice patch."[11]

In the Alps, the ice melt is exposing gruesome artifacts from World War I:

> The glaciers of the Italian Alps are slowly melting to reveal horrors from the Great War, preserved for nearly a century.
>
> In the decades that followed the armistice, the world warmed up and the glaciers began to retreat, revealing the debris of the White War. The material that, beginning in the 1990s, began to flood out of the mountains was remarkably well preserved. It included a love letter, addressed to Maria and never sent, and an ode to a louse, 'friend of my long days', scribbled on a page of an Austrian soldier's diary.
>
> The bodies, when they came, were often mummified. The two soldiers interred last September were blond, blue-eyed Austrians aged 17 and 18 years old, who died on the

---

[11] Julie Cruikshank, *Do Glaciers Listen?: Local Knowledge, Colonial Encounters and Social Imagination* (2005), 245.

> Presena glacier and were buried by their comrades, top-to-toe, in a crevasse.[12]

While buried in a crevasse, pretty clearly that they are now exposed reveals how much the ice is returning to World War I levels. On other continents, melting glaciers are exposing other recent reminders of where man used to go:

> On the other side of the world, glaciers in the Argentinian Andes have relinquished their grip on a different set of bodies: Incan children sacrificed five hundred years ago, and a young pilot who crashed just a few decades ago.
>
> "It took me a very long time to acknowledge he might be dead," the pilot's mother said, reported Stephen Messenger for Treehugger in 2011. "Now we have a body. I can visit my son at his burial site and grieve like any mother has a right to do."
>
> A different plane carrying 52 passengers crashed into an Alaskan glacier in 1952. An Alaska National Guard helicopter crew found the wreckage in 2012.[13]
>
> The soldiers at Matienzo Base in Antarctica made an odd discovery in January 1995—an antique dog sled, unlike any they had ever seen, bound together by leather straps, with a label reading "Made in England."

---

[12] Laura Spinney, "Melting glaciers in northern Italy reveal corpses of WW1 soldiers," *Telegraph*, January 13, 2014.
[13] Marissa Fessenden, "As Glaciers Retreat, They Give up the Bodies and Artifacts They Swallowed," *Smithsonian*, May 27, 2015.

> Matienzo was an Argentine research base on a small island 30 miles off the coast of the Antarctic Peninsula. It was surrounded by a vast plain of white—a slab of glacial ice, 700 feet (215 meters) thick, that floated on the ocean. That floating slab, called Larsen A Ice Shelf, covered an area of water the size of San Francisco Bay. It had existed for a thousand years or more.[14]

While the article portrayed this as more evidence of global warming, the fact that artifacts a century old are now being exposed might have caused the question: "Why is this stuff appearing *now*? Have we only returned to the ice conditions of a century ago?"

> Around A.D. 1400, what is now western Canada entered a minor ice age. After this cooling peaked around 1700, glaciers started to recede in the 1850's, said Dr. Blake, who is president of Icefield Instruments Inc., a glacier research company in Whitehorse. In recent years, the process accelerated, converting areas that elderly Indians remember as ice patches into expanses of moss-covered rock.[15]

Of course, these might be indications that glaciers grew during the Little Ice Age, and are now melting and exposing the very recent past. But we have other reminders that man has long lived and worked at what are now melting glacier

---

[14] Douglas Fox, "Scientists Are Watching in Horror as Ice Collapses," *Scientific American*, April 12, 2016.

[15] James Brooke, "Lost Worlds Rediscovered as Canadian Glaciers Melt," *New York Times,* October 5, 1999.

altitudes:

> Only a few sites in the Alps have produced archaeological finds from melting ice. To date, prehistoric finds from four sites dating from the Neolithic period, the Bronze Age, and the Iron Age have been recovered from small ice patches (Schnidejoch, Lötschenpass, Tisenjoch, and Gemsbichl/Rieserferner).... These finds date from the Neolithic period, the Early Bronze Age, the Iron Age, Roman times, and the Middle Ages, spanning a period of 6000 years. The Schnidejoch, at an altitude of 2756 m asl[Above Sea Level], is a pass in the Wildhorn region of the western Bernese Alps. It has yielded some of the earliest evidence of Neolithic human activity at high altitude in the Alps. The abundant assemblage of finds contains a number of unique artifacts, mainly from organic materials like leather, wood, bark, and fibers. The site clearly proves access to high-mountain areas as early as the 5th millennium BC, and the chronological distribution of the finds indicates that the Schnidejoch pass was used mainly during periods when glaciers were retreating.[16]

Neolithic finds of leather, bark and wood have been found that "can be attributed to three time slots between 4800 and 2200 BC...."[17]

In Canada:

---

[16] Albert Hafner, "Archaeological Discoveries on Schnidejoch and at Other Ice Sites in the European Alps," 65 *Arctic* SUPPL. 1 (2012) 189.
[17] *Ibid.*, 193.

> High in the Mackenzie Mountains, scientists are finding a treasure trove of ancient hunting tools being revealed as warming temperatures melt patches of ice that have been in place for thousands of years....
>
> Ice patches are accumulations of annual snow that, until recently, remained frozen all year. For millennia, caribou seeking relief from summer heat and insects have made their way to ice patches where they bed down until cooler temperatures prevail. Hunters noticed caribou were, in effect, marooned on these ice islands and took advantage.
>
> "I'm never surprised at the brilliance of ancient hunters anymore. I feel stupid that we didn't find this sooner," says Andrews.
>
> Ice patch archeology is a recent phenomenon that began in Yukon. In 1997, sheep hunters discovered a 4,300-year-old dart shaft in caribou dung that had become exposed as the ice receded. Scientists who investigated the site found layers of caribou dung buried between annual deposits of ice. They also discovered a repository of well-preserved artifacts.[18]

In Colorado:

> To the untrained eye, University of Colorado at

---

[18] Arctic Institute of North America. "Ancient artifacts revealed as northern ice patches melt." *ScienceDaily*. www.sciencedaily.com/releases/2010/04.

> Boulder Research Associate Craig Lee's recent discovery of a 10,000-year-old wooden hunting weapon might look like a small branch that blew off a tree in a windstorm.
>
> Nothing could be further from the truth, according to Lee, a research associate with CU-Boulder's Institute of Arctic and Alpine Research who found the atlatl dart, a spear-like hunting weapon, melting out of an ice patch high in the Rocky Mountains close to Yellowstone National Park.
>
> Lee, a specialist in the emerging field of ice patch archaeology, said the dart had been frozen in the ice patch for 10 millennia and that climate change has increased global temperatures and accelerated melting of permanent ice fields, exposing organic materials that have long been entombed in the ice.[19]

In Norway:

> Mittens, shoes, weapons, walking sticks – lost in the high mountains of Norway thousands of years ago - are now emerging from melting ice....
>
> Around 7,000 years ago the Earth was enjoying a warm climate. Now glaciers and patches of perennial ice in the high mountains of Southern

---

[19] University of Colorado at Boulder. "Hunting weapon 10,000 years old found in melting ice patch." *ScienceDaily*. www.sciencedaily.com/releases/2010/06.

Norway have started to melt again, revealing ancient layers.

"Actually we should be slowly approaching a new ice age. But in the past 20 years we have witnessed artefacts turning up in summer from increasingly deeper layers of the glaciers," says Lars Pilø.

He is an archaeologist working for Oppland County, and has for many years done fieldwork in glaciers and ice patches, finding things our ancestors discarded or lost.

The summer of 2014 was hectic in this respect. In Oppland County alone, Pilø and his colleagues found 400 objects, now emerged from the deep freeze.

Among these were a horse skull and hiking staffs from the Viking Age. An arrow shaft found by the archaeologists is from the Stone Age.[20]

Similarly, remains of Native Americans have emerged from melting glaciers in North America in 1999.[21] Perhaps the best known such ancient body is that of Ötzi the Ice Man, defrosted in 1991 in the Alps.[22]

Yellowstone ice patches are also exposing human artifacts:

---

[20] Marianne Nordahl, "Items lost in the Stone Age are found in melting glaciers," *Science Nordic*, January 16, 2015.

[21] Robert James Muckle, *Indigenous Peoples of North America: A Concise Anthropological Overview* (2012), 51.

[22] Jorge Daniel Taillant, *Glaciers: The Politics of Ice* (2015), 210.

Ice patches don't move like the larger glaciers, so they are even better suited for preserving material. In the greater Yellowstone ecosystem — an area that includes surrounding mountains and forests — researchers started collecting these artifacts about eight years ago, Craig Lee, an archeologist at the Institute of Arctic Alpine Research in Colorado, told Dayton.

In Yellowstone, Lee, archeologist Staffan Peterson, and others have found animal bones, wooden weapons, and other artifacts ranging from 10,000 years to just a few hundred years old. They've carbon-dated leaves and tree stumps that are more than 5,000 years old. [23]

Are sea levels rising? It's not the first time.

During the last Ice Age, so much water was locked up in glaciers that sea levels 20,000 years ago were 120 meters (about 370 feet) lower than today:

---

[23] Marissa Fessenden, "Melting Ice in Yellowstone is Revealing Ancient Artifacts Faster Than Researchers Can Handle," *Smithsonian*, September 3, 2015.

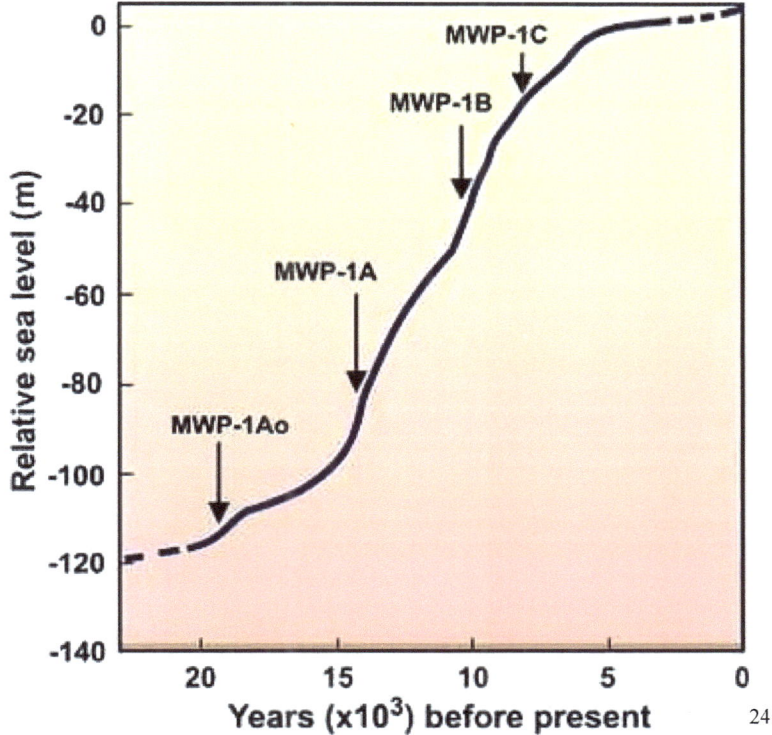

The East Coast used to be out to sea from the current coastline because of this.

---

[24] Vivien Gornitz, "Sea Level Rise, After the Ice Melted and Today," NASA Science Briefs, January 2007.

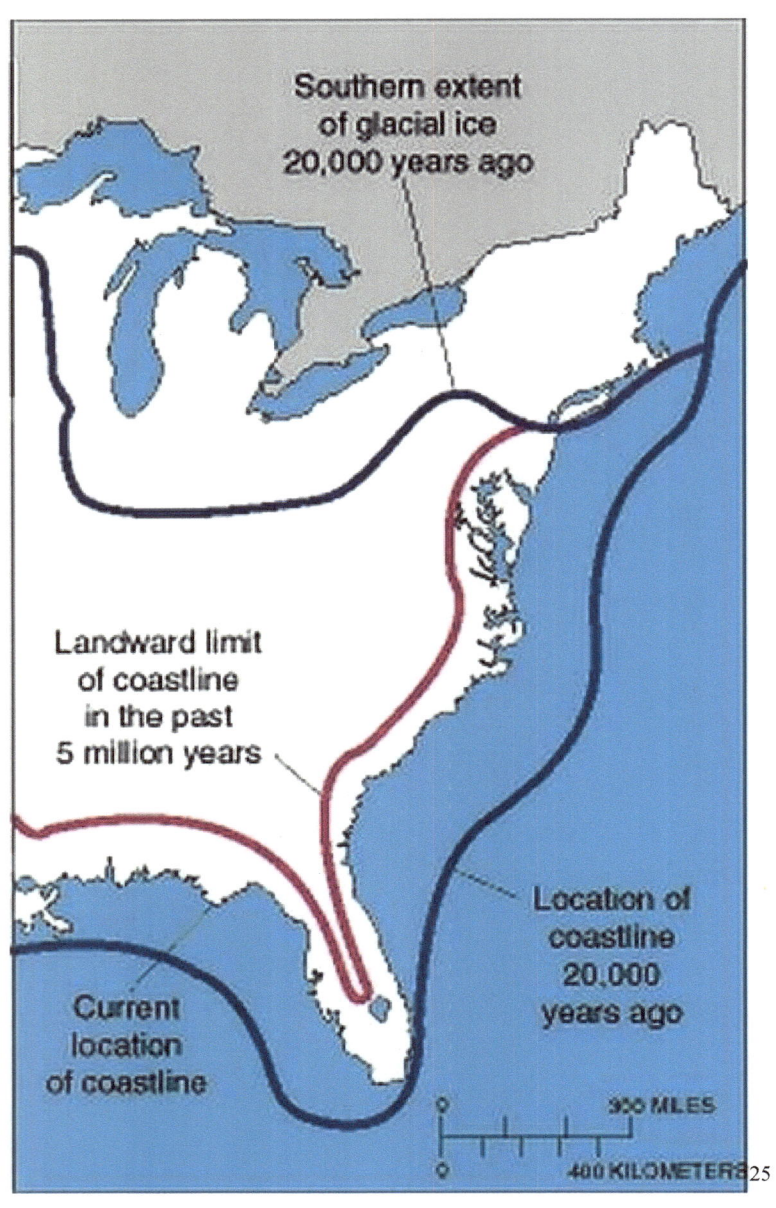

[25] United States Geological Survey, Sea Level, http://pubs.usgs.gov/circ/c1075/sea.html, last accessed May 7, 2016.

Other coastlines were similarly affected. On the following map: blue areas were covered by glaciers and green areas were dry land then and below sea level now. Areas outside the current shorelines on this map were above sea level:

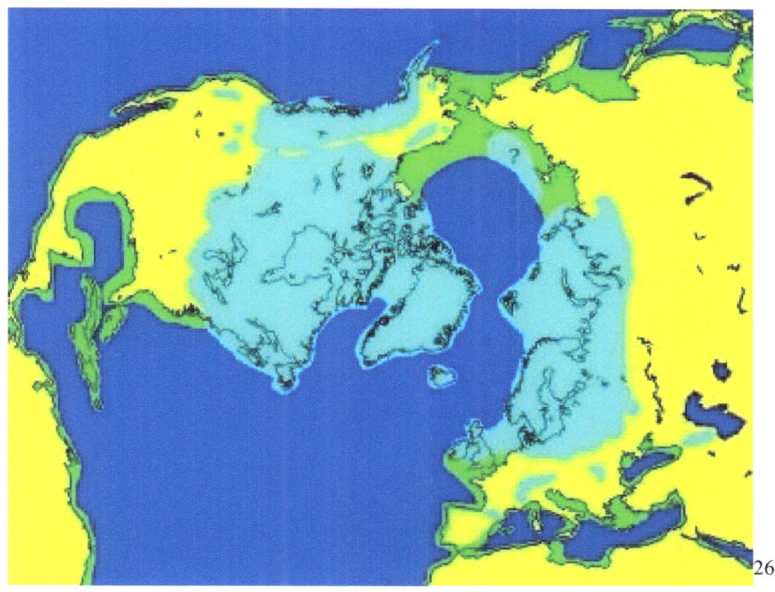

Along with changing sea levels, changing rainfall patterns altered many lake shorelines. When drought hit Switzerland in 1857 and again in 1921, falling water levels in Lake of Morat and Lake of Neuchatel exposed Iron Age settlements, showing that humans had settled both areas in the past.[27]

---

[26] Steve Dutch, Pleistocene Glaciers and Geography, University of Wisconsin Green Bay, https://www.uwgb.edu/dutchs/EarthSC202Notes/GLACgeog.HTM, last accessed May 7, 2016.

[27] H.F. Westlake, "Notes on Some Recent Excavations at Westminster Abbey," *Antiquaries Journal* 1:237 (1921).

4 Not the First Time the Glaciers Are Melting

It is an article of faith that glaciers melting away are the result of man's use of fossil fuels, which really took off in the 20th century, as warmists insist:

> Here we show that the anthropogenic component (atmospheric value reduced by the pre-industrial value of 280 ppm) of atmospheric carbon dioxide has been increasing exponentially with a doubling time of about 30 years since the beginning of the industrial revolution.[28]

But "the glaciers are melting!" is hardly a new observation. At the close of the 19th century, it was generally recognized that whether the world's glaciers were advancing or in retreat was a serious question, as this 1892 article from *American Geologist* acknowledged.

> *Are the Glaciers of North America Advancing or Retreating?*
>
> The glaciers of this continent have been known for so short a time that only small portions of their histories have been read. Their study is comprised almost entirely within the past decade and has been carried on in such a desultory way that for the most part only qualitative evidence

---

[28] David J. Hofmann, James H. Butler, Pieter P. Tans, "A new look at atmospheric carbon dioxide," *Atmospheric Environment* 43:12 2084-2086.

as to their advance or retreat is available.[29]

And in this book about the glaciers on Mt. Rainer in Washington State:

> RECESSION AND SHRINKAGE OF THE GLACIERS.
>
> Every glacier about Mount Rainier that was examined by the writer furnished evidence of a recent recession of its terminus and of a lowering of its surface. In two instances—the Carbon and Willis glaciers— rough measurements of the amount of these changes during the past fifteen years were obtained.... The forests are advancing on the barren areas and gradually taking possession of them. This evidence, even if actual observations of the recession of the extremities of the glaciers were not available, is sufficient to show that the ice streams have for a number of years been growing shorter and shorter.[30]

And this 1899 article about global melting of the glaciers:

> If the glaciers of the world are becoming smaller and showing a prevailing tendency to retreat, or in other words, to move a lesser distance down the valleys, the fact is of much interest in its bearing upon climatic changes. Professor

---

[29] Israel Cook Russell, "Climatic Changes Indicated by the Glaciers of North America," *American Geologist* 9:324 (May, 1892).
[30] Israel Cook Russell and George Otis Smith, *Glaciers of Mount Rainier* (1898), 407.

Russell, of Michigan University, has shown that the minor advances and retreats of glaciers may be due to causes that are not meteorological, but the fact remains that changes in the larger glacial movements can be explained only by variations in the quantity of snow received and in the rate of melting caused by climatic fluctuations. The fact that glaciers are subject to quite rapid variations in volume, and that the tendency, in recent years at least, has seemed to be toward reduction in volume in all parts of the world, has excited much interest. The International Geological Congress at Zurich, in 1894, appointed a committee to collect data from different quarters of the globe with regard to the variation in the size of glaciers. Inquiry in this direction was greatly stimulated. The Alpine Clubs have been active, and most of the European glaciers have been systematically studied. A great deal of information has also been gathered from Central Asia and North America.

The "Deutsche Rundschau," for Geography and Statistics, has recently given a summary of the latest reports on the retreat and advance of glaciers, from which it appears that in the Swiss Alps thirty-nine glaciers are receding, five are stationary and twelve are advancing. The glaciers in the corner of Bavaria that push into the Alps are all receding, and also those of the Hollenthal in Paden and the Sonnblick group in the eastern part of the Austrian Alps. None of the Italian glaciers are advancing, while many are receding. The Cassandra group has recently retreated about eighty feet, and one glacier in the Bernina group has receded 3,508 feet in seven years. One of the Swedish glaciers has retreated 393 feet, and the glaciers in Norway are also receding. The recent studies of Spitzbergen glaciers show that some of them have retreated more than a mile and a half; but it is not known, of course, how long this recession has been going on. In America many glaciers have receded to the snow line, as the limit of perpetual snow is called. The remarkable report comes from Turkestan not only that are the glaciers receding but also that some of them have entirely disappeared, and a similar report comes from the Altai Mountains on the southern edge of Siberia.[31]

The *Alpine Journal* in 1878 reported on the subject at great length:

I Presume that one of the objects of the 'Alpine Journal' must be to contain a record of Alpine changes; and, as no one else has undertaken the task, I have thrown together some details of the very remarkable retreat of the glaciers which has now continued for some years, hoping that in the absence of anything better, this record, though imperfect, may be serviceable hereafter, and may also be not unacceptable to the readers of the ' Alpine Journal.'

---

[31] *Current Literature: A Magazine of Contemporary Record* 26:6 (December, 1899), 549.

> Mr. F. F. Tuckett, in the sixth vol. of the 'Alpine Journal,' p. 30, has given an exceedingly interesting account of the retreat of the Lower Grindelwald Glacier, and of the uncovering of ancient marble quarries from beneath it. He also gives reasons for believing that the glaciers were at a minimum about the year 1750; that between that time and 1771 they had rapidly increased, and had not again diminished until after 1850, since which time the retreat has been continuous and very remarkable. Up to about that time the glaciers were no doubt increasing, and the Swiss expressed some anxiety on the subject. Soon afterwards the advance ceased, and in many of them a retreat began; not, however, in all of them, for the Findeln Glacier at Zermatt was in 1859 encroaching on the pastures and turning up old turf like a gigantic ploughshare. The natives give as a reason for the retreat, an unusual number of mild winters, and most old Swiss travellers will say that the summers have on the whole been hotter and finer than they were formerly. The inhabitants of Chamonix declare that their crops ripen a fortnight earlier, and attribute this to the diminution of the glaciers; more, however, is probably due to the warm summers and to improved cultivation.[32]

The *Journal of Geology* in 1908 also reported on the melt:

> *French Alps.*—The glaciers in the Grandes

---

[32] C. Marett, "On The Retreat Of The Swiss GLACIERS," *Alpine Journal* 8:275 (1878).

Rousses of Dauphine are in general retreat. A map of these glaciers on a scale of 1/10000 is now being made. Measures of snowfall in the Savoy have shown a smaller amount in the winter of 1904-5, than in that of 1903-4. Special observations on the Mont Blanc chain have shown that the greatest snowfall occurs at an altitude in the neighborhood of 2,550 meters. The glaciers of Mont Blanc and the Maurienne show a slight retreat though there are indications of increased activity which later may bring on an advance. In the Vanoise and the upper valley of the Arc the glaciers continue to retreat, and some large snow fields have disappeared; others have been broken up by projecting ridges of rock.

Pyrenees.—The glaciers in these mountains are stationary or retreating. There have been very great changes since the middle of the last century; for instance, between 1855 and 1904, the glacier de l'Est has retreated 1,140 meters and the glacier de la Breche, 1,230 meters. In the last two years there seems to be an increase of snowfall on these glaciers. The disappearance of some small glaciers in the French Alps and in the Pyrenees has been injurious to agriculture on account of the decreased quantity of water available for irrigation. This has led the Minister of Agriculture to offer pecuniary support to glacial observations.

Sweden and Norway.—One glacier was

observed in Sweden in 1905, the Mika, and it has retreated three to four meters. In Norway the changes have been mixed, some glaciers have retreated and some have advanced. The three glaciers observed in the Jostedal have advanced from 5 to 19 meters.

Russia.—In the mountain chain of Peter the Great, Boukhara, two glaciers show an advance since 1899, one of them as much as 64 meters. One in the Tian-Chan shows a retreat since 1892.

Caucasus.—Many glaciers have been visited and named; the Bartui has steadily been retreating; the retreat amounted to 30 meters in 1900-1, 12 meters in 1902-3, 13.5 meters in 1903-4. The glaciers of the Caucasus seem to be in general retreat.

British Columbia and Alberta.—The Illecillewaet glacier continues to retreat, but much more slowly; it lost but 2 feet 6 inches between 1905 and 1906, though there has been a general shrinking in the volume of the ice. The tongue of the Asulkan glacier is slowly melting away under the moraine.

South America.—A short description of the glaciers of Poto, just north of Lake Titicaca, Peru, has been given by Otto F. Pfordte. The San Francisco glacier has high terminal moraines, but the present end has not varied much since the Spanish occupation, as shown by the ruins of houses at the foot of the cliff, where the glacier now ends. Old observations and traditions of the natives indicate that the snowline is gradually receding in this part of the Andes, which accounts for the gradual lowering of the lakes. Mr. Bandelier, referring to this same general neighborhood, states that the glaciers of the Bolivian Andes have been in slow retrocession for a number of years.

Central Africa.—The Mubuhu glacier on the eastern slopes of Ruwenzori is apparently in retreat. An old moraine overgrown with vegetation may be recognized some 500 meters in advance of the existing tongue of the glacier, and from the appearance of the rocks nearby it would seem that a slow retreat is now in progress (1905). Morainic lakes have been observed on the western slope below the limits of the present glaciers by Dr. Stuhlmann.

REPORTS ON THE GLACIERS OF THE UNITED STATES FOR 1906

The snow fall in the Rocky Mountains in the summer of 1905 was very heavy and perhaps for that reason the Hallet Glacier shows a slight advance *(Mills)*. But there has been a slight retreat at the north end of Arapahoe Glacier, which is not far from the Hallet *(Henderson)*. The glaciers in the Montana Rockies are either stationary or slightly retreating *(Chaney)*. A small glacier reported in Bighorn Mountains of

Wyoming has apparently disappeared *(Salisbury)*.

On the north side of Mt. Hood, Washington, Eliot Glacier is diminishing very markedly. The ice is growing much thinner at the end and a more rapid retreat will probably appear before long. Some of the snowfields are greatly altered and the ascent of the mountain has been rendered much more difficult than heretofore *(Mrs. Langille)*. On the south side of Mt. Hood, the White Glacier has diminished in thickness but does not seem to have receded materially *(Montgomery)*. Glacier Peak, in Washington, was climbed last summer by Mr. C. E. Rusk. He found that the glaciers showed signs of retreat but less than in other places in Washington; these glaciers carry comparatively little debris. …

Last summer Messrs. F. E. and C. W. Wright visited Glacier Bay and repeated the survey which was made in that region in 1892. They found very remarkable changes in all the glaciers. The only definite information we have had of any of these glaciers since 1899, until the Messrs. Wright's visit, was due to a trip to Muir Glacier by Messrs. Andrews and Case, in May, 1903, and they reported a very considerable recession in that glacier. The report of the Messrs. Wright will not be published before next winter but they have very kindly prepared an abstract of the glacial changes which they observed as follows;

On comparing our map with your map of 1892, the following changes are most apparent: Beginning with Muir Glacier and its tributaries the ice front has receded a maximum distance of about 33,000 feet; Dirt Glacier is no longer tidal; White and Adams Glaciers are supplying very little ice to the general ice field; Morse Glacier terminus is about one mile from tide water; the crest of the stagnant ice mass between Girdled Glacier and Muir Inlet has melted down about 200 feet since the time of your measurements; Girdled Glacier and Berg Lake, however, have not changed materially in aspect. The length of the total ice front of Muir Glacier is now over 40,000 feet instead of 9,000 feet in 1892. The present ice front passes at its northern extremity at about the position of your 1,000 feet contour on the ice of 1892. This remarkable decrease in elevation is undoubtedly due not only to melting down but also to breaking down of the exposed ice masses. The ascent of the ice mass at this point is decidedly steep and the ice fairly cascades into the water. The present height of the ice fronts of all the tide water glaciers is about the same as noted by you in 1892 (150'-250'), and is a noteworthy fact in connection with these glaciers. Muir Inlet is at present choked by the ice pack which promises to remain congested so long as its source of supply is so active. A considerable portion of the present front of Muir Glacier is in very shallow water and in a few years should decrease in size very materially unless new avenues and inlets for tidal currents are exposed by the receding ice. Dying Glacier is still creeping back and wasting away.

Carroll Glacier has not changed much in aspect during the last 14 years; its terminal cliff has receded about 2,000 feet and at present, apparently, is continuing to do so. It is discharging icebergs very slowly and Queen Inlet is nearly free of ice.[33]

---

[33] Harry Fielding Reid, "The Variations of Glaciers," *Journal of Geology* 16:49-53 (1908).

4 Chronicles

A common subject of 19th century and early 20th century writing concerning climate change was the regurgitation of chronicles of extreme cold in the past. While historians regard chronicles as not very meaningful history ("just one damn thing after another," without analysis, as one of my undergraduate professors described them), one must presume that if an event was considered important enough to record, it probably happened. The next few pages may well cause you to use that phrase; instead of reading every line and year, look at the remarkable incidents such as the Black Sea and Adriatic Sea freezing.

From *The Scrap Book* (1907):

> A Review of Famous Drops of the Temperature Indicates That " Oldest Inhabitants" Are Right When They Assert That Winters of the Olden Time Had More Vigor and Snap Than Those of To-day.
>
> Is the earth getting warmer or colder? Science seems rather inclined to answer that it is getting warmer, and the scientific view finds support from numerous "oldest inhabitants," who, especially in the United States, [assert] that the winters of the last two or three decades lack the snap and vigor of the winters of the olden time, when fence tops were concealed by great snow blankets that were from four to five feet thick. But here are some cold weather records which will aid the readers of THE SCRAP BOOK in

solving the vexed question for themselves.

401. Black Sea was frozen over for twenty days.

462. The Danube was frozen so that an army crossed on the ice.

463. A frost in Constantinople which began in October, and continued until February.

768. The Black Sea and Strait of the Dardanelles were frozen over.

822. The Danube, the Elbe and Seine were frozen so hard as to bear heavy wagons over them for a month.

860. The Adriatic was frozen.

874. Snow fell from the beginning of November to the end of March.

891 and 893. Nearly all vines in Europe were killed by frost.

1035. A frost in England on Midsummer's Day was so severe that it destroyed fruits.

1133. The Po was frozen from Cremona to the sea. Wine-casks were burst, and trees were split by the action of the frost.

1216. The Po was frozen. Wine casks were burst.

1234. Loaded wagons crossed the Adriatic to Venice.

1236. The Danube was frozen to the bottom, and remained so for a long time.

1261. The Kattegat was frozen from Norway to Jutland.

1292. The Rhine was crossed by loaded wagons, and travelers crossed the ice from Norway to Jutland.

1323. Foot and horse travelers crossed from Denmark to Lubeck and Danzig.

1344. All the rivers of Italy were frozen over.

1403. The wolves were driven by the cold from Denmark, and crossed the ice to Jutland.

1434. It snowed forty days without interruption.

1460. The Danube was frozen for two months.

1468. The wine distributed to the soldiers in Flanders was cut in pieces with hatchets.

1544. The same thing happened again, the wine being frozen in solid lumps.

1565. The Scheldt was frozen so hard as to bear loaded wagons over it for three months.

1594. The Adriatic was frozen at Venice.

1621 and 1622. All the rivers of Europe were frozen; the Hellespont was covered with a sheet of ice, and the Venetian fleet was frozen up in the lagoons of the Adriatic.

1658. Charles X of Sweden crossed the Little Belt, the strait between Funen and the peninsula of Jutland, with his whole army—foot, horse, baggage and artillery. The rivers in Italy bore heavy carriages.

1664. The Thames, in England, was covered with ice sixty-one inches thick.

1684. Coaches drove across the Thames.

1691. The cold was so intense that the wolves entered Vienna, and attacked men and cattle in the streets.

1693. Again famished wolves attacked men and beasts in the streets of Vienna.

1695. Many people were frozen to death in Germany.

1709. This was that famous winter called by distinction, "the cold winter." All the rivers and lakes were frozen, and even the sea for several miles from the shore. The ground was frozen in England nine feet deep. Birds and beasts died in the fields, and men perished by thousands in their houses. In the south of France, the olive-trees were killed, and the wine plantations mostly destroyed. The Adriatic Sea was frozen, and even the Mediterranean about Genoa. The citron and orange groves suffered extremely in Italy.

1716. The winter was so intense that persons traveled across the straits from Copenhagen to Sweden. Fairs were held on the river Thames.

1726. In Scotland, multitudes of cattle and sheep were buried in the snow.

1737. In January the ground in New England was frozen four feet deep.

1740. An ox was roasted whole upon the Thames. The winter was scarcely less cold than that of 1709. The snow lay ten feet deep in Spain and Portugal. The Zuyder Zee was frozen over, and thousands of people went over it. The lakes in England also were frozen. During the hard frost, a palace was built of ice at St. Petersburg, after an elegant model, and in the just proportions of Augustan architecture.

1744. Snow fell in Portugal to the depth of twenty-three feet on the level. This was a summer winter in New England.

1754 and 1755. These winters were very severe. In England the strongest ale exposed to the air in glass was covered with ice one-eighth of an inch thick. These were very mild winters in New England.

1771. The Elbe was frozen to the bottom.

1774 and 1775. These winters were very severe. The Little Belt was again frozen over. On January 11, 1774, the thermometer in Portland, Maine, was fourteen degrees below zero, and on the 22d at the bottom of the plate.

1776. The Danube bore ice five feet thick below Vienna. Vast numbers of birds and fishes perished. In Holland and France wine froze in the cellar.

1796. Perhaps the coldest day ever known in London was December 25, 1796, when the thermometer was sixteen degrees below zero.

1800, January 13. Quicksilver was frozen hard at Moscow. [That would be at least -40° C.]

From 1800 to 1812. The winters were remarkably cold, particularly the latter in Russia, which surpassed in intensity that of any winter in that country for many preceding years, and caused the destruction of the French army in its retreat from Moscow. What with the loss in battle, and the effects of this awful and calamitous frost, France lost in the campaign of this year more than four hundred thousand men.

1848. January 10, at eight o'clock, the

thermometer in Portland was three degrees below zero. January 11, at same place, thermometer fifteen to twenty degrees below zero. At Boston eleven degrees below. In some parts of New England the thermometer stood thirty-nine and forty degrees [below zero]; and the day was known as the " Cold Tuesday."

1858. The winter, so mild in the United States, was particularly severe in Europe. For the first time in the nineteenth century, the Po was frozen over at Ferrara, admitting for a long period the constant passage of man and beast. At Constantinople, on February 3, snow fell without interruption for fifteen days. There had not been a winter of equal severity for more than twenty years. The snow extended to Smyrna, and the adjacent districts of Asia Minor and the Greek Islands were clothed in white, an appearance unusual and remarkable. In Asia Minor a Greek monastery was buried, and five monks had to be excavated by the Turks. At Malta, snow, which had not fallen since the year 1812, was several feet high, and accompanied with hail and tempests. The navigation with Odessa was closed.

There are no points in Europe where the cold records of America are eclipsed, but in Asia our lowest records are thrown completely into the shade. Siberia has the coldest weather known anywhere in the world. At Verkhoyansk, Siberia, 90.4 degrees below zero was observed in January, 1888, which gets away below anything ever known in the world before or since. At that point the average temperature for January is nearly 64 degrees below. This town is situated at an elevation of three hundred and thirty feet above the level of the sea and during the entire winter the weather is nearly always calm and clear.

Not the least remarkable of cold weather records is "the year without a summer"—1816. In that year, there was sharp frost in every month, and people all over the world began to believe that some great and definite change in the earth was taking place. The farmers used to refer to it as "eighteen hundred and starve to death." January was mild, as was also February, with the exception of a few days. The greater part of March was cold and boisterous. April opened warm, but grew colder as it advanced, ending with snow and ice and winter cold. In May ice formed half an inch thick, buds and flowers were frozen and corn killed. Frost, ice, and snow were common in June. Almost every green thing was killed, and the fruit was nearly all destroyed. Snow fell to the depth of three inches in New York and Massachusetts, and ten inches in Maine. July was accompanied with frost and ice. On the 5th ice was formed of the thickness of window-glass in New York, New England, and Pennsylvania, and corn was nearly all destroyed in certain sections. In August ice formed half an inch thick. A cold northern wind prevailed nearly all summer. The first two weeks of September were mild, the rest of the month was cold, with frost, and ice formed a quarter of an inch thick. October was more than usually cold, with frost and ice. November was cold and blustering, with snow enough for good sleighing. December was quite mild and comfortable.[34]

Another list often reprinted in the early 19th century:

In A. D. 400, the Black Sea was entirely frozen, as was the Rhone in all its length. Such a phenomenon indicated a temperature of at least 18° centigrade, 4° Fahrenheit, below zero. When the Gulf of Venice was frozen in 1709, the thermometer in that city fell to 20° centigrade, 4° Fahrenheit, below zero.

"In A. D. 462, the army of Theodomer crossed the Danube on the ice. The Var, a small river of France and Italy, falling into the Mediterranean, between Nice and Monaco, was frozen, which effect demanded a temperature of 10° or 12° centigrade, below zero.

A. D. 763, the Black Sea and Dardanelles were frozen.

A. D. 822, loaded carriages traversed on the ice for upwards of a month, the Danube, the Elbe, and the Seine. The Rhone, the Po, and

---

[34] *The Scrap Book* (1907), 614-16.

the Adriatic sea were frozen. See A. D. 400.

A. D. 829, on the authority of Abd Allatif, translated by M. Silvestre de Sacy, when the Jacobite Patriarch of Antioch, Dionysius of Telmahre, attended the Khaliffe Mamoun into Egypt, they found the Nile frozen.

A. D. 860, the Adriatic sea and the Rhone were frozen, demanding a temperature of 20° centigrade, 4° Fahrenheit, below zero.

A. D. 1133, the Po was closed from Cremona to the sea, and the Rhone crossed on the ice. Wine froze in the cellars:—at least 18° centigrade, 4° Fahrenheit, below zero.

A. D. 1216, the Po and the Rhone frozen; and again in 1234, the same rivers were closed, and loaded carriages traversed the Adriatic sea on the ice near Venice. (20° centigrade, 4° Fahrenheit, below zero)

A. D. 1236, the Danube closed for some considerable time.

A. D. 1292, loaded carriages crossed the Rhine below Breisach, and the Kattegat sound completely closed.

A. D. 1302, Rhone frozen (—18° cent., 4° Fahrenheit below zero.)

A. D. 1305, the Rhone, and all the other rivers of France were frozen.

A. D. 1323, the Rhone frozen. Travellers on foot and horseback passed on the ice from Denmark to Lubeck and Danzig.

A. D. 1358, ten feet of snow at Bologna in Italy.

A. D. 1364, the Rhone frozen to Ariles; loaded carriages passed on the ice. -18° cent., -4° Faht.

A. D. 1408, the Danube frozen in all its course; one sheet of ice from Norway to Denmark. Carriages crossed the Seine on the ice.

A. D. 1434, frost commenced at Paris, the last of December, 1433, and continued during three months, less 9 days; recommenced towards the end of March and continued to the 17th of April. The same year it snowed in Holland 40 consecutive days.

A. D. 1460, the Danube and the Rhone frozen.

A. D. 1493, the port of Genoa frozen.

A. D. 1507, the port of Marseilles frozen in all its extent. (-18° cent., at least, -4° Faht.) On the day of Epiphany, 3 feet of snow fell at the same city.

A. D. 1468, the wine had been reduced to ice and cut with an axe; and in 1544, a similar severity of cold in France.

A. D. 1565, the Rhone was frozen to Arles. (-

18° cent., -4° Faht.)

A. D. 1568, from the 11th to the 21st of December, the Rhone passed on the ice. (—18° cent, at least.)

The winter of 1570-1571, from the end of November to the end of February, was so severe, that all the rivers, even those of Languedoc and Provence, were so completely frozen that they were passed with loaded carriages. (Mezerai.)

A. D. 1594, the sea at Marseilles and Venice frozen. (-20° cent., -4° Faht)

A. D. 1603, loaded carriages passed the Rhone on the ice. (- 18° cent., -4° Faht.)

The winter of 1621-1622, the Venetian fleet arrested by the ice in the lagoons of Venice; in 1638, a similar event with the French galleys at Marseilles; either event demanding a temperature of -20° cent, or - 4° Fahrenheit.

(A. D. 1645, the Swedish army passed from Holstein into Zealand on the ice.)

In the winter of 1655-1656, the Seine was closed from the 8th to the 18th of December. It was again frozen, without interruption, from the 29th of December to the 28th of January. A new frost recurred a few days after, and continued until in March. (Bouillaud.) The ensuing winter,

1657-1658, an uninterrupted frost from the 24th of December, 1657, to the 8th of February, 1658. Between the 24th of December and the 20th of January the cold was moderate, but afterwards acquired an extreme intensity. The Seine was entirely closed. A slight thaw took place on the 8th of February, but the frost again recurred and continued to the 18th. It was in 1658, that Charles X, king of Sweden, traversed the Little Belt with his army, artillery, caissons, baggage, &c.

A. D. 1662-1663. Intense frost at Paris, from the 5th of December to the 8th of March.

A.D. 1676-1677, continued and very intense frost from the 2d of December to the 13th of January; the Seine was closed 35 consecutive days.

A. D. 1684, the Thames, at London, frozen 11 inches thick, and traversed by loaded wagons.

A. D. 1709, (perhaps the most intense season which has ever occurred within the range of history,) the Adriatic sea, and the Mediterranean from Genoa by Marseilles to Cette, frozen. All the rivers and narrow seas of Europe frozen. (-20° cent., -4° Faht.)

A. D. 1716, booths erected on the Thames at London.

A. D. 1726, sledges passed from Copenhagen to

> Sweden.

> A. D. 1740, the Thames, at London, again frozen.

> From 1749 to 1781, (33 years,) the thermometer, in Provence, never fell below —9° cent. (20° Faht). This period of 33 years, afforded no instance of a cold of from 15° to 18° below zero, as formerly; some persons already concluded that the climate had meliorated; but in 1789, this illusion was dissipated, because in that year, they experienced at Marseilles, a cold of -17° cent. -4° Faht.[35]

Both of these collections appear to come from The Tablet of Memory (1783).[36]

Nineteenth century accounts also provide evidence that the advance and recession of glaciers was known:

> In the polar regions, they say, many passages, traversed by navigators at no very remote period, are now rendered impassable, by the ices which obstruct them. The some effects may, according to them, be observed in our most elevated mountains, where the glaciers are perceived every century, and almost every year, to descend towards their bases. We are assured that they cannot fail, by this slow, but certain progress, finally to invade our fields, meadows, and villages.

---

[35] William Darby, *A View of the United States, Historical, Geographical, and Statistical* ...(1828), 415-19.

[36] *The Tablet of Memory, Shewing Every Memorable Event in History*, 5th ed. (1783), 76-77.

> With respect to the Alps, particularly those of Switzerland, it is certain that the ice upon them has, in the course of some latter years, been very perceptibly extended.
>
> In the bailiwick of Interlaken, the snows now occupy some parts of the mountains which formerly afforded pasturage, and they have entirely obstructed a ward which led beyond them into le Valais. A small village, called Sainte Petrovelle, has disappeared, and the ground upon which it stood is covered with ice.[37]

Other accounts tell of the end of Petrovelle:

> The upper glacier, situated between the Wetterhorn and the Mettenberg, is 1 1/2 league in length, full of rifts and crevasses, and has very pure ice pyramids of all kinds. The lower glacier, situated between Mettenberg and the Eiger, is the more interesting. It is a sea of ice, three leagues in length, terribly torn and cleft, and scattered with pyramids of a grotesque form. "Scarcely three hundred years ago an open pass, several leagues in length, led over the chain into the Valais, from which people came t« the church of Grindelwald to celebrate baptisms and weddings. To-day all is covered with a wild and impassable sea of ice "— ZSCBOKKE.

---

[37] *Kaleidoscope, Or, Literary and Scientific Mirror* (June 14, 1825), 5:259.

> In the seventeenth century the glacier increased in an extraordinary manner, and was no longer to be contained by its valley. Bursting its barriers, it carried away the dwellings which lay in its course, and destroyed the church of Petrovelle, the bell of which, cast in 1044, is yet in the church of Grindelwald. Tradition relates that at one time the Mettenberg and Eigers formed but one mass, behind which was a lake of considerable size.[38]

More recent works acknowledge the overrunning of Swiss villages by growing glaciers:

> Historical records of the latest glacier fluctuations are available from many other alpine regions in Europe. In the European Alps, the oldest records involve the Great Aletsch Glacier in Switzerland. During the thirteenth century, this glacier advanced over part of an aqueduct used to transport meltwater to a local village. More recent records throughout the Alps chronicle glacier advances (and recessions) over the past few centuries. Villages built in what had been considered safe places were overwhelmed by glaciers in the early seventeenth century. Several of these former villages are still ice covered today. Many distinct advances characterize this period of glacier expansion, major ones culminating in the early 1600s, the 1820s, and the 1850s. A glacial recession, interrupted by a minor readvance

---

[38] David Bogue, *Switzerland and Savoy* (1852), 54.

between 1912 and 1920, characterized the period from 1850 to ~1960.[39]

The same work also describes clear evidence of glacier fluctuations over human history associated with warmer climates in Roman times, and increased cold in the eighteenth century.

---

[39] John P. Bluemle, Joseph M. Sabel, and Wibjörn Karlén, "Rate and Magnitude of Past Global Climate Change," *Geological Perspectives of Global Climate Change: AAPG Studies in Geology* (2001), 47:203.

5 What Does This Mean?

Pretty clearly, Earth has experienced great variations in temperature in the last several thousand years, with dramatic advances and recessions of the world's glaciers. The advances are demonstrated by historical reports of glaciers overrunning existing settled lands; the recessions are demonstrated by the emergence of artifacts and remains as ice returns to its old levels. The reports of receding glaciers now just over a century old show that this is not the first time, and cannot be blamed on carbon dioxide that has entered the atmosphere in the last few decades. To argue that we are in an extraordinary change because of 20th century fossil fuel use requires ignoring a lot of historical and archaeological data.

ABOUT THE AUTHOR

Clayton E. Cramer teaches history at College of Western Idaho. His MA in History is from Sonoma State University in California. His scholarly work has been cited in two U.S. Supreme Court decisions, *DC* v. *Heller* (2008) and *McDonald* v. *Chicago* (2010) as well as dozens of lower court decisions. He lives in Horseshoe Bend, Idaho.

www.ingramcontent.com/pod-product-compliance
Lightning Source LLC
Chambersburg PA
CBHW040918180526
45159CB00002BA/522